The Philosopher's Stone Book

STEVEN SCHOOL

DISCLAIMER

This publication is intended for informational purposes only. Neither the author nor the publisher assume any liability for the use or misuse of the information contained herein. No warranty is expressed or implied as to the accuracy or completeness of any information presented here. This book does not constitute advice of any type, nor is it intended for any specific person. Do not try any of these things at home. Do not consume any substances. Always consult a qualified physician for medical advice.

DEDICATION

This written work is dedicated to the modern generation of inquisitive minds and is influenced by the hand of time. It is an alchemical tract upon the great work of the sun and moon or the separation and conjunction thereof in due proportion as is done in accordance with nature.

CONTENTS

ACKNOWLEDGMENTS

As the great and venerable father of lights has told us in the emerald tablets, It has its birth in the earth, the wind (water) has carried it in its belly, its strength it doth aquire in the fire, and from this the one thing, come all things by adaptation.

Salt To The Cross.
S.A.S. 2016.

www.howtomakethephilosophersstone.com

1 INTRODUCTION

In the ancient world of alchemy there were two kinds of people, those who knew the secrets of the art and those who did not. These two classes of persons were described in the bible as the ignorant and the wise and this was also symbolized by the awakening of Adam and Eve when they consumed the forbidden fruit of the tree of knowledge of good and evil. It has been written that the shepherds tend to their flocks of sheep, being those who are forbidden to partake of such secret knowledge in order to keep the separation of classes for if everyone were equal then there would be no kings or queens to rule over the lower world. Throughout history there have been secret meetings of secret societies marked by symbolism which is found everywhere. A secret cup, a secret drink, drink brother, and live was the motto of the initiated ones. Jesus at the last supper, holding up a wooden cup, the holy grail for all to see but understood only by the wise. The chosen few or the illuminated ones. The ancient science covered a great many topics such as medicine, science, metallurgy, mathematics, astrology, astronomy and more. Hermes Trismegistus was called the father of science and was credited with being a key figure in the further development of the hermetic art. The ancient Egyptians utilized the ankh as their symbol for eternal life because they believed that man was intended to live forever in perfect health without sickness or death. This theory is marked by the tree of life which is written of in the bible. There are some who believe the mighty oak tree can live for thousands of years and further that since God created all things equal to grow and to multiply in like kind, that so it should also be with us and with all other things including the metals and the stones. Eternal life marked by the tree of life and symbolized by a secret garden called Eden for the chosen few who found the way or were otherwise initiated, the illuminated ones who walk the earth as "Gods" considering themselves to be more than just mortals simply because they possess knowledge which has been withheld from others for thousands of years. Jesus was said to have been a carpenter, and most everyone knows that they work with wood. He was also said to have travelled the land miraculously healing the sick with a quantity of whitish colored powder. The primitive alchemical process began with a simple formula of fire and water to act upon matter. This was also seen when various Indian tribes built canoes in which they would select a fallen tree and use fire to hollow it out before quenching it with water. They would then scrape out the charred portion and repeat this work until the canoe was shaped and ready to use. They found it much easier to cut the wood with fire than with the hand tools of a common workman and this is alchemy, the ancient formula of fire and water. These are interesting points to consider as we progress throughout the rest of this book.

Steven School. 2016.

2 ANCIENT MEDICINES

The tree of life.

Ancient alchemists believed that diseases and sicknesses of the body were only a side effect or a symptom of an imbalance of the individuals ph, while issues involving the mind were associated with ammonia in the brain or the bloodstream. They also believed in one medicine, a universal medicine which would neutralize acid or even ammonia and bring us back to an alkaline ph balance so that the body could heal or repair itself by generating new cells. This "medicine" was said to cause a strengthening of the limbs (bones), and was also said to be known by the fact that it causes plants to flourish. They believed that perhaps we were never meant to wither and die but instead to continue growing like the mighty oak tree, here in the garden that was built for us. Over the years I have heard stories of near death experiences which included brilliant white lights and tales of glory and splendor. I have news, when I was a child of approximately five or six years old my grandmother took me on a road trip to Tehachapi because she wanted to look at land for sale in the hopes of building her dream home for her retirement. To make a long story short I will get right to the point of the matter. As she met with the sales personnel I was left at the playground which had one of those tall metal slides typical of the early to middle nineteen seventies. An older kid knocked me off of the slide and I landed on my back on the sand, I hit the back of my head on the concrete footer for one of the upright supports. The world began to spin and then everything faded to black. I woke up three days later in the hospital and my grandmother was sitting by my bed. She said I had gotten a concussion from hitting my head on the concrete, but when I landed on my back my heart had stopped. She told me that by the time the paramedics arrived my heart was not beating, I had no pulse, I also was not breathing. I was completely unresponsive and they informed her that I was dead. My grandmother was hysterical, they tried everything they could, and they managed to do some good it seems because I did wake up three days later. Many years passed and I thought back to that time, remembering what had occurred. I even began to describe the events to others whenever I heard people talking about the persons on TV describing the afterlife or near death experiences and so forth. According to what I went through my understanding is that I have been to the other side and come back. What I saw was nothing, blackness, emptiness, a complete lack of existence. That time is gone, there was nothing there which brought me to the realization that if we are to find the eternal life which is promised to us in the bible that it must come before death and not after since death is the opposite of life. Everything that we have in death, is exactly the opposite of what we had in life, yin and yang, white and black, light, and darkness. The eternal sleep of death, or the gift of everlasting life. Alchemists had an interest in the mighty golden oak. For its strength, its longevity, and its continual growth. The golden oak tree, the golden soma.

One morning I awoke and prepared to go to work, I noticed something different on this day, my knees hurt and they felt like bone against bone. The joints did not want to work correctly and I could hear clicking noises when I tried to get up or down which was also quite difficult. This had come on quickly and was unexpected. I began to worry, would I be crippled? Would I be able to function and to take care of myself? This prompted me to research the matter online and the first thing I came across during an internet search which caught my attention is that achy joints and especially the knees is a sign of an improperly functioning liver. I knew that when I was born my body created what it needed, bones, cartilage, vital organs, brain matter, etc. I quickly realized that when my liver was not functioning properly, it stopped my body's ability to regenerate and to repair itself as nature had intended. My research indicated that the liver could supposedly regenerate new cells to repair itself in a three month period. I put down the alcoholic beverages, I drank ice water with fresh sliced lemon. I went to two different vitamin stores to get supplements as well as ordering some online which they did not carry. I began with milk thistle pills which were supposed to be good for my liver, I also chose shark cartilage pills, fish oil capsules, and Echinacea herbal tea. I started to ride my bicycle again as well. First one lap around the block, then two, then three… My knees feel great now. I have been hearing about others who chose surgery instead which can leave scar tissue. I put my faith in mother nature first and she didn't let me down. The moral of the story is this, I hypothesize that my body is meant to heal itself! My arthritic knees were only a side effect of an underlying issue! I almost forgot to mention one of the supplements that I bought and it is one of my utmost favorites, coral calcium which is rumored to help oxygenate the body on top of being a great source of calcium in my opinion. Oxygen… the breath of God! When I consider biblical accounts of people supposedly living for one thousand years or more I contemplate the fact that both the air and the water quality must have been so much better in their time. No thousands of automobiles stuck in rush hour traffic burning up my precious oxygen supply, no fluoride and birth control being literally pumped to my faucets. And then there is the biblical writings that instruct us not to eat leavened bread, leaven means yeast which is a living organism that feeds on sugar to create alcohol. I believe the bible is right about not wanting this in our body. It also says not to eat cloven hoofed swine, microorganisms?, parasites?, worms? I also would like to mention something that I discovered recently, both potatoes and tomatoes are a member of the nightshade family of plants. Nightshade is poisonous. The potatoes and tomatoes however are only very mildly toxic but because of this many natural healers advise not to eat them, no more French fries with ketchup, mashed potatoes, potato salad, etc. I developed varicose veins prematurely in life part of this I am certain is due to receiving a third degree burn but not all of it. I have been an avid coffee drinker for many, many years now. I may drink it morning, noon, evening or even night. One pot of coffee is enough for me at breakfast time. I decided to quit drinking it but after six hours my mind and body said dude, to the hell no! I felt like my brain had shrunk, it apparently now is a sponge for caffeine. After all of these many years of over indulging it is proving a hard habit to break. My research indicates that blood vessels are not resilient, I do not believe that they have any elasticity to them which mean if they are stretched out, they do not return back to their original size or shape. Coffee contains caffeine which gets the blood pumping full speed ahead buddy, but what happens when the effect wears off? Are my blood vessels left loose and stretched out?, I think so. If this hypothesis is correct then would it not adversely affect my cardiovascular system? At least the caffeine is pumping my coral calcium supplements throughout my body. Being that I am currently single I eat mostly microwaveable prepackaged frozen stuff. This has come to my attention because I keep getting little growths on the back of my head. Cancer comes to mind and for some reason my instinct tells me to consider the microwave. Now, let us get back to ancient medicine. So the alchemists from long ago were said to have believed in a universal medicine, a golden elixir, a golden soma. The biblical tree of life comes to my mind here, where is this thing?, what is this thing? Let us begin with the first word of its description, tree. Like a slap in the face could it be that simple? The ancient sages wrote about their golden bough or their golden branch, as well as a golden soma, or a golden elixir. In their riddles they loved to dance around and hint at the oak tree. One in particular in my mind, the golden oak tree. I scooped ashes from my fireplace, (oak ashes), I ground them to powder and baked them using a casserole dish in my oven. My intent was to purify the ashes in heat by burning away any combustible impurities. I placed the cooled matter into my coffee pot with a few filters stacked up and brewed it just like coffee. The water which filled the pot was of a golden color, I evaporated some of it to dryness and was left with a white powder. The alkaline salt of potash is an interesting topic when we delve into the writings which lay ahead in this section. The ancient alchemists warned that too much (overconsumption) of their secret "elixir" would fire the body and exhaust the spirit. My own personal hypothesis is that too much potassium could probably cause a heart attack. I did notice that when I sprinkle ashes into my garden it seems to be the best plant food that I have ever seen, it causes the vegetation in my yard to flourish, lush and green. I sprinkle around wood ashes and then wait for mother nature to bring the rain. Rainwater and ashes causing my plants to flourish. Two thousand years ago in the first century Pliny the elder wrote Historia Naturalis which I believe means natural history. Two thousand years takes us way back into the depths of alchemy. What a great place to dig for insights into the ancient science! The writings of course are seemingly never ending but yielded a gem. In those times, Pliny suggested that one might let thy hearth be thy medicine chest. A hearth is a fireplace and what does it contain but wood ashes? Archeologists have uncovered old gladiator bones from the roman era. While studying the remains to determine what their diet may have been, it was determined that they drank a medicinal beverage of ashes from the fire pit mixed with water. I believe this is also high in strontium. Reports indicate that this drink helped speed recovery from wounds and their bones were also reported to have been stronger or denser than those of regular people of the time. I recall that Jesus supposedly walked the land healing the sick, he was said to have been a carpenter and they work with wood. Some people believe that he had a bag of white powder that he added to water, (turned the water into wine). I have heard some opinions that the holy grail is Jesus cup, and that it was supposedly made of wood. I believe that in the image of the last supper he may be holding up such a cup for the world to see. Wood, fire, and water, a drink, a medicine, alchemy. Perhaps a secret meant only for those who have eyes to see? Let's take a look at what Moses has to say, wasn't he supposed to have lived for about 986 years or so?

EXODUS 32:20 ENGLISH STANDARD VERSION.

He took the calf that they had made and burned it with fire and ground it to powder and scattered it on the water and made the people of Israel drink it.

I believe that long ago, in the forgotten era before video games were invented, that some people used to carve figurines out of wood.

The salt of the world?, the salt of the earth?.

Matthew 5:13King James Version (KJV)

13 Ye are the salt of the earth: but if the salt have lost his savour, wherewith shall it be salted? it is thenceforth good for nothing, but to be cast out, and to be trodden under the foot of men.

John 4:13-14King James Version (KJV)

[13] Jesus answered and said unto her, Whosoever drinketh of this water shall thirst again:

[14] But whosoever drinketh of the water that I shall give him shall never thirst; but the water that I shall give him shall be in him a well of water springing up into everlasting life.

I would like to mention now my opinion on the tree of knowledge of good and evil. That tree from which Adam and Eve were said to have partaken of the forbidden fruit. Forbidden, outlawed, banned, illegal, persecuted, prosecuted, **expelled from the garden baby, hands off.**

Genesis 2:16-17King James Version (KJV)

[16] **And the LORD God commanded the man, saying, Of every tree of the garden thou mayest freely eat:**

[17] **But of the tree of the knowledge of good and evil, thou shalt not eat of it: for in the day that thou eatest thereof thou shalt surely die.**

I am going to share my understanding of this matter in simple terms, Cannabis is not a plant, it is a tree. I have seen the trees big and tall, and with bark on them. What plant grows eighteen or more feet tall with thick bark on it? A tree. Researchers are now theorizing that cannabis causes neurogenesis which is the body's ability to repair its own damaged brain by growing new cells. Reminds me of my liver and my knees which we covered earlier. Consumption of the "forbidden fruit" seems to stimulate deep and profound thought. There are some persons out there who hypothesize that this material may have healing qualities towards things like cancer. It has also been rumored that this substance might have the ability to repair brain damage caused by excessive alcohol consumption. Let us progress now, to the next subject that I would like to cover.

Throughout history vinegar has been used as a medicinal tonic often infused with such things as herbs, spices, essential oils, garlic, onions, turmeric or a wide variety of other things. It has been used topically as well as internally. I drink a tiny amount once in a while diluted in ice water, I also sometimes use a little bit of apple cider vinegar topically on my psoriasis. Another home remedy that I have tried is a little baking soda in a glass of water. I hypothesize that it might be alkalizing or perhaps balancing the PH. I further surmise that it may neutralize ammonia in the bloodstream which of course is only my thoughts or opinion and does not constitute advice of any type.

Ancient Greek practitioners of medicine such as Hippocrates (400 B.C.) were said to have mixed apple cider vinegar with honey as a medicament for a variety of ailments. Vinegar was also supposedly used around 218 B.C. to crumble large boulders. A fire was built against the large rocks to get them very hot and then the vinegar was poured on causing the boulders to crack. Water and fire, alchemy at work, I hope they wore their safety glasses. I believe we have covered Cleopatra dissolving pearls in vinegar in the section about alchemical gem stones. There have been rumors that vinegar may be useful in the reduction or elimination of microorganisms. During the time of Jesus vinegar was also called wine which can be seen in the bible and this is interesting because it may help to understand certain verses from that book. During medieval times vinegar was infused with garlic and consumed as a medicinal beverage to ward off the plague. In modern times this is supposedly called four thieves vinegar. Vinegar has been used in the past as an antiseptic to clean and disinfect wounds. The European alchemists of the middle ages were also known to have used vinegar in their alchemical works regarding the philosopher's stone.

As a tree grows soluble minerals and nutrients are carried up into it by the action of water where they theoretically become locked within the wood. Alchemists believed that these building blocks of nature could be released and separated through the action of fire and water. From blackness comes whiteness, the white dove.

3 THE SECRET FIRE

In Researching the history of alchemy one tends to come across references to a secret water that was believed to be required in order to perform or conduct the great work of the magnum opus. This substance was rumored to contain what the alchemists called the secret fire. In the writings of Theophrastus Paracelsus he suggested that this water was sold by the apothecaries of his time. John Pontanus wrote that he had failed more than two hundred attempts at the creation of his stone until he read the written alchemical works of Artephius which he credited for giving him the proper understanding of the matter. So what is this seemingly elusive water?

From the writings of Artephius, **ARGENT VIVE.**

Alchemists loved to communicate through symbolism, secret codes, and anagrams such as argent vive. Simply rearrange the letters to reveal the secret.... VINEGARET. **Vinegar** in modern terminology.

In Nicholas Flamels letter to his nephew he mentioned his advice on this subject, **(know with what agent your "mercury" must be fortified with or it will be as common water).**

White vinegar is mostly distilled water with a small amount of acetic acid. The acetic acid is the "secret fire" contained in the water which was required in order to perform the alchemical magnum opus. In modern times this is simply called the metal acetate path.

The secret key which unlocks the metals.

4 THE PHILOSOPHER'S STONE

The term Philosopher's stone sounds to most people as if it infers one secret and mystical stone, while yet others still believe that perhaps it was even mythical in nature. We shall begin this section with an illumination of what the "stone" was. Alchemy is a study and or replication of nature. The simple and ancient method of fire and water acting upon matter. Alchemists knew three basic areas of work, the plant, animal, and mineral realms. Medicines for mammals were said to be found in the first two kingdoms while tinctures for minerals such as metals and gem stones were believed to be found in the latter. The method of working in the mineral kingdom has been called in modern times the metal acetate path. Metallic ores were worked upon by the ancient sages with vinegar to produce toxic metal acetates which were further processed into what were hypothetically called philosopher's stones. Since there is more than one metallic ore that would be compatible with the metal acetate path, there was more than one philosophers stone. There were as many different stones as there are such compatible ores. Each "stone" had its own color spectrum according to the mineral content of the ore. Some ores might be harder to break down so they might have been more compatible with the dry path which began with roasting. I feel it is important to note here even though this section is not about techniques or methods however roasting ores produced what was called the poisonous breath of the dragon which slays or kills everything in its path. Do not try any of these things at home, do not breathe any fumes, do not consume any substances. This book is written for historical reference purposes only and is not meant to constitute advice of any type. So theoretically speaking there could be as many different philosophers stones as there are metallic ores compatible with the metal acetate path. Alchemists invented dyes for many things such as glass, fabrics, dishes, plates, cups, goblets, tapestries, and according to legend metals as well as gem stones. Each stone had its own color spectrum as we have mentioned previously. An example of this would be red for iron (Mars) while iron and sulfur (Iron Pyrite) is associated with the color of gold. According to alchemical belief the alchemist assisted nature in the creation of their stones, the materials worked upon were chosen by color spectrum according to the intent of each individual artist. (What they intended to use their stone for). And the basic idea was that these provided color for alchemical gem stones as well as transmutation (amalgamation) of metals. There are some who believe that when nature creates gemstones within the earth's crust that the color comes from broken down or decomposed metallic ores. This is interesting because many hard rock gold miners believe that gold is often found in Limonite veins wherein Iron Pyrite crystals have decomposed. So then perhaps the practitioners of the ancient science intended to follow the work of nature in creating and or coloring metals and gems. Another belief was that all things descend or evolve towards gold over time and this is interesting when I look at pyritized fossils. Pyrite suns, (the alchemical sun sounds familiar here) pyrite snails, pyrite eggs, etc. decomposed pyrite crystals in limonite veins, gold.

Some persons like to think of the stone as a salt crystal, and to compare the work to basic crystal growing.

This would appear to simplify the matter.

5 THE GUALDUS WET PATH

Trituration- To grind into a fine powder, as fine as the painters grind the colors. Credit- Theophrastus Paracelsus.

The sealed microcosm of the alchemist. In modern terminology this might be called an ecosystem. The matter was ground to powder and placed into the retort (one part). The vinegar was added (two parts). Alchemists liked to begin the great work in spring and progress through the summer months in accordance with nature so that no external heat was needed. Room temperature or sunlight for a solar distillation. As Flamel said, the warmth of a hatching chicken. In winter months some alchemists buried their vessel under their house in the dirt when using the one vessel method, others used fresh horse dung, warm ashes, even lye to keep the glass warm or close to body temperature. The work proceeded slowly and naturally, dissolving, extracting, subliming, circulating, exalting, distilling. The agent and the patient, the volatile and the fixed.

As the vinegar dissolved matter in the retort it began to release the naturally occurring sulfuric acid in the iron pyrite. This clear liquid was called the blood of the green lion (Iron sulfide) and was gently distilled over the helm with the white vinegar by the hand of nature, alchemists warned that the practitioner only sets the proper conditions, nature does the work, without the laying on of hands. In the retort occurred the color changes as the work progressed. Black, white, yellow, the peacocks tail, and red.

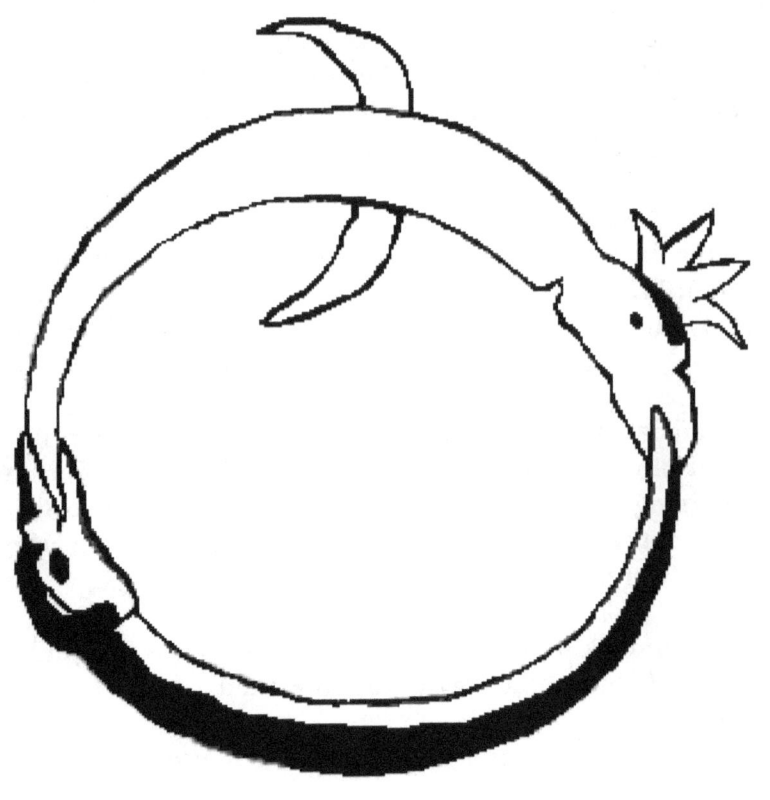

What the Ourobos means, the fixed iron pyrite in the vessel below, the volatile vinegar leaving the matter and going over the helm of the retort, it is in a circle because it will be returning over and over again. When the dry land appears, (the pyrite is dry) the vinegar in the receptacle is poured back upon the iron pyrite. Each time this happened completed one turn of the alchemical wheel. With each repetition the vinegar takes more sulfuric acid from the matter being dissolved, this multiplication or exaltation (circulation) was continued until all of the "gold" (sulfuric acid) went over the helm. "mercury" of seven eagles was said to sway the moon (produce the white stone), "mercury" of ten eagles was said to have power to calcine the sun, (finish exalting the pyrite into the philosopher's stone). When the vinegar had taken the sulfuric acid over the helm into the receptacle the ancient alchemists then called it "our most sharp vinegar", or "well actuated mercury".

Actuated= activated. The liquid became stronger or more powerful with each turn of the alchemical wheel. "Burning" or "calcining" the matter by "water" not fire. Hence the term alchemists burn with water not fire. A philosophical calcination in the "wet path".

This Ourobos represents the great work of the sun and moon, King and Queen, the volatile and the fixed.

Each circulation supposedly exalted the matter further.

6 THE SENDIVOGIUS METHOD

One vessel. Wet path.

The matter was ground to powder and placed into the vessel. The vinegar was added and the top covered with a breathable dust cover to let evaporation occur while keeping insects or dust out. the vinegar dissolves, extracts and sublimes the matter. In this type of alchemical sublimation the dissolved matter rises in the liquid and adheres to the sides of the glass in the upper portion while the impurities fall to the bottom of the jar. At dryness the iron pyrite was wetted again with fresh vinegar and this process repeated eleven times. The first matter of metals (Flamels mercurial sublimate or the white stone) hypothetically stuck to the glass first, in the latter imbibitions the fixed salt (alchemical seed of gold) was finally released from the broken down ore. The two mingled in the water during the final imbibitions leaving the philosopher's "stone" stuck to the upper portions of the jar where it could be scraped off after being allowed to dry. There was said to be another step after the mercurial sublimate or "virgins milk" was collected and it was called inceration which was to "fix" the matter and to render it fusible like wax so that it would withstand the fire, and this was done in heat. Now let us understand this in Sendivogius words from the new chemical light.

The first matter of metals is twofold, and one without the other cannot create a metal. The first and principal substance is the moisture of air mingled with warmth. This substance the Sages have called Mercury, and in the philosophical sea it is governed by the rays of the Sun and the Moon. The second substance is the dry heat of the earth, which is called Sulphur.

Its appearance is that of oily water adhering to all pure and impure things; yet in some places it is found more abundantly than in others because the earth is more open and porous in one place than in another, and has a greater magnetic force. When it becomes manifest, it is clothed in a certain vesture, especially in places where it has nothing to cling to. It is known by the fact that it is composed of three principles; but, as a metallic substance it is only one without any visible sign of conjunction, except that which may be called its vesture or shadow, sulphur.

The metals are produced in this way: after the four elements have projected their power and virtues to the center of the earth, they are, in the hands of the archeus (water) of nature then distilled and sublimed by the heat of perpetual motion towards the surface of the earth. For the earth is porous, and the air by distillation through the pores of the earth is resolved into a water out of which all things are generated. (archeus is vinegar).

The Artist only separates what is subtle from its grosser elements and puts it into the proper vessel. Nature does the rest. Out of one arise two, and out of two arise one.

INCERATION.

The "virgins milk" that is expressed from the better part of the stone is then carefully preserved in an oval shaped distilling vessel made of glass and is day by day wondrously changed by the quickening fire.

Credit, Michael Sendivogius.

This concludes the Sendivogius wet path.

7 THE FLAMEL DRY PATH

In the wet path of alchemy which we have already examined the alchemist's first cooked their "fire" in their "water" and then later roasted the matter which was called inceration. The dry path of alchemy is the same however the steps were simply reversed and it was also said to be much faster. The dry path was believed to be more dangerous since the alchemist's were roasting their ores, while the longer wet method supposedly produced a better final product. During the roasting of the ore the color changes occurred showing all of the colors of the peacocks tail until the final fixed red was achieved. The fire broke down the matter and burned away the combustible impurities. This resulted in the red lion which was then further processed by placing it either in the retort (Gualdus method) or the jar (Sendivogius method) and then imbibed with the vinegar. The ancient alchemist's then proceeded with the imbibitions or circulations until the exaltation of the matter was complete. Theophrastus Paracelsus preferred the alembic for the alchemical magnum opus (wet or dry methods). So to simplify this, the dry path was the same as the wet path except the matter was thoroughly roasted first. During the circulations color changes were seen again.

Nicholas Flamel was believed to have discovered the secrets of alchemy after a lifetime of diligent study, it has also been said that even with the secret knowledge he remained a humble book seller and was known for donating large sums to charities including churches, hospitals, and housing for the homeless. His tomb was rumored to have been found empty.

8 METALLIC TRANSMUTATION

Metallic transmutation of metals has been contemplated by researchers for centuries. Some have pondered nuclear fusion while others have considered cold fusion. Scientists have hypothesized that elemental sulfur is the nucleus of the gold atom, some have expressed their opinion that when metals are produced naturally in active lava flows eight times more gold might be produced if sulfur is present in the equation. The ancient alchemists experimented with the idea of breaking down the metals to extract their salt and sulfur principles using philosophical "mercury" (vinegar). One theory is that perhaps these salt and sulfur principles were to be joined or fused together to create a stone. I believe that transmutation is old terminology and that in this modern era we might simplify the matter by calling it amalgamation. In primitive metallurgy potash was used as a fluxing agent to purify metals as well as for amalgamation. Wood ash was calcined and ground to powder. This material was mixed with metallic ores in crucibles and smelted before being poured into molds and allowed to cool. The resulting piece of metal was then knocked loose from the mold and the slag chipped away. This process was believed to cleanse the metal by separating the impurities into the potash which solidified on top. This appears to be the basis which lead to the invention of steel (an exalted form of iron). Once the metal was cleansed of its impurities it was ready for amalgamation during which more of the flux could be added. My understanding is that the metal would have then been smelted again in a crucible with the fluxing agent over a wood fire, then the molten mass stirred with an iron rod while dropping the "stone" into the mix. The stirring continued until the desired effect was achieved and then poured into molds and allowed to cool usually in the shape of bars. Small indents were scratched into the ground to serve as makeshift molds and the resulting amalgam were called finger bars. These were metal bars small like a finger and hence the name.

The athanor was the furnace of the alchemists. Even the ashes were useful for different purposes as we have seen in this book.

9 ALCHEMICAL GEMSTONES

In my alchemical works or studies I began to experiment into the calcination of oak wood. I have a wood burning fire place in which I try to use only wood so that my ashes are free from contaminants. The last fire had been long gone and I scooped out some of the charred oak ashes. I placed this material into mason jars with lids to keep it clean for my studies. I purchased a new casserole dish with a lid for about fifteen dollars at my local store and then I ground some of the ash to a fine powder in one of my glass mortar and pestles. I placed this material into the dish and baked it in my oven for a couple of hours at around 300 or more degrees. I turned the oven off and went to bed. A couple of days later I baked it for another couple of hours, I repeated this procedure a few times and increased the heat each time until I was baking at the highest temperature that my natural gas burning oven would make. A couple hours here, a couple hours there, increasing the heat. One day I removed the cooled lid to see what I had, I was expecting to see light grey well calcined ashes…. However when I first collected my ashes some of them were black chunks of charred wood, which I had ground to a fine powder, now I once again had some chunks of black material looking like it had returned to the condition it had been in before it was ground to powder… interesting. There was a difference however, these chunks were shaped like squares and rectangles and reminded me of large cut gem stones due to the sizes and shapes however they looked like charred black lumps. I decided I would grind these again in my mortar and pestle, they were very, and I mean very, hard to break. I feared that my mortar and pestle would break first however I finally managed to crack one of the pieces which was much harder than wood. I began to contemplate, wood, ashes, charred, charcoal, carbon, heat… and then it dawned upon me. The ancient alchemists were rumored to have the ability to create large gem stones of exquisite beauty. And then at that very

25

moment it made perfect sense how they had made the discovery, so simple, by accident really. In this study of nature the secrets just seem to fall into the possession of the diligent pursuer. Such a simple discovery. The writings of Theophrastus Paracelsus offer an insight as well into the coloring of alchemical stones. Metallic bhasmas, extracts from metallic ores, yes the philosophers stones from the caverns of the metals and exalted by the hands of men. Pervading with color, beautiful hues of blue, green, azul, fire like that of gold imparted into a clear stone reminding me of topaz, the brilliance of the diamond, the beautiful red of the ruby tinged by iron (Flamels God of war), and the sheer elegance of the emerald. The ancients were also believed to have the ability to dissolve pearls with the intent to use the resulting tincture to create larger or more valuable pearls. Here is a bit of the goody that I found in my research which fits nicely here. The queen of Egypt Cleopatra was said to have dissolved pearls in vinegar before consuming a portion of the resulting tincture which she believed to have medicinal qualities or some type of health benefit. This gives a good portion right here of how the ancients might have begun a work of creating alchemical pearls.

10 THEORY OF TIME TRAVEL

Time is measured as the earth rotates on its axis. One revolution basically equates to 24 hours or one day. As this occurs the earth also rotates around the sun which is the center of our universe in a counter clockwise direction. In this fashion time is moving forward. In one year light can travel roughly 6 trillion miles which equals one light year. Earth years and light years are measured differently and so to travel in space is to travel in time. Since the earth rotates counter clockwise, if a craft or "object" were to orbit the earth in the same direction while travelling at the speed of light it would theoretically be travelling into the future. If the craft were to reverse direction this would be considered travelling back into the past. Another interesting point to consider is that sometimes aircraft fly from one time zone to another, imagine leaving tonight and arriving yesterday morning, now multiply that by one hundred million times over by increasing speed.

Steven and Belle.

MATHEW 5:13

[13] Ye are the salt of the earth: but if the salt have lost his savour, wherewith shall it be salted? it is thenceforth good for nothing, but to be cast out, and to be trodden under foot of men.
[14] Ye are the light of the world. A city that is set on an hill cannot be hid.
[15] Neither do men light a candle, and put it under a bushel, but on a candlestick; and it giveth light unto all that are in the house.

ABOUT THE AUTHOR

Some have asked the question, if you discovered the knowledge of alchemy why would you share it with the world and not just keep it for yourself?

Proverbs 3:16
Blessed is he who finds wisdom;
For she is more precious than pearls;
And nothing that you desire compares with her;
Length of days is in her right hand;
And in her left hand are riches and honor;
All her ways are pleasant;
And all her paths are peace;
Behold, Dianna unveiled.

S.A.S. 2016.

www.howtomakethephilosophersstone.com

www.ingramcontent.com/pod-product-compliance
Lightning Source LLC
Chambersburg PA
CBHW021448170526
45164CB00001B/435